# ADAS Colour Atlas of Weed Seedlings

**John B Williams**
**John R Morrison**

Wolfe Publishing Limited are proud to be linked with the Agricultural Development and Advisory Service (ADAS) in this publication.

# ADAS Colour Atlas of

# WEED SEEDLINGS

## John B Williams

ADAS Horticultural Publications Officer
Luddington Experimental Horticulture
Station
Stratford-upon-Avon

## John R Morrison

Principal Photographer
Harpenden Laboratory
Hatching Green
Harpenden

## Cathy Wood
Illustrator

A Wolfe Science Book

© Crown Copyright 1987. Published by permission of the Controller of Her Majesty's Stationery Office
Published by Wolfe Publishing Ltd 1987
Printed by Royal Smeets Offset b.v., Weert, The Netherlands
ISBN 07234 0929 3

This book is one of the titles in a rapidly expanding science series. For further information on books in the series, plus forthcoming titles, please write to Wolfe Publishing Ltd, Wolfe House, 3 Conway Street, London W1P 6HE.

# Contents

# Acknowledgements

Much help and encouragement have been given by Mr P M Dawson, ADAS Publications Officer; by Mr H A Roberts and Mrs J E Boddrell, National Vegetable Research Station. Mr Roberts kindly provided the data on the emergence of seeds from the soil.

All the photographs of seedlings (except those of common hemp-nettle) are by John Morrison.

Photographic prints by Mrs P Moody, Harpenden Laboratory.

Line illustrations by Cathy Wood.

Photographs of adult plants (exceptions listed below) are by John Williams.

Photographs of common hemp-nettle were kindly supplied by ICI plc.

Photographs of flowering plants of hairy tare and pale persicaria supplied by The Weed Research Organization; of scarlet pimpernel, parsley-piert and corn marigold by Mr W Bond, National Vegetable Research Station.

# Introduction

For the professional farmer, grower and consultant, the ability to recognize weeds at the earliest two-leaf (cotyledon) and young plant stage can give significant cost savings in herbicides. Weeds vary in their susceptibility to different herbicides and this susceptibility varies with stage of growth. Identification at an early stage will allow you to choose the cheapest spray to clean the crop effectively.

For the amateur gardener, weed identification is born of curiosity although it is helpful to be able to recognize the weeds when cleaning the flower and shrub borders. Many of the farm weeds are common in gardens but several are common in gardens and rare in the fields; these too are illustrated.

Pages 10 to 13 show miniature photographs of the common weeds 'at a glance'. These, together with the notes on identification and the identification drawings appearing at the top of left-hand pages, should help the reader to find the right group of illustrations quickly.

A glossary of terms used in the descriptions appears on page 95.

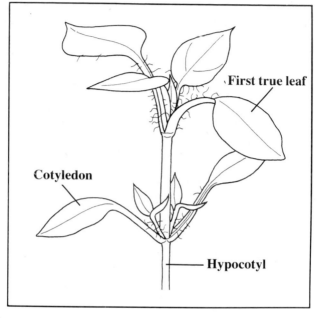

Cotyledon

First true leaf

Hypocotyl

For a full glossary see page 95

# Five questions about a weed seedling

**What do you notice first?**

1 Is it large?
2 Is it very small?
3 Is the hypocotyl long; are the cotyledons carried proud of the soil or do they lie close to it? What colour is the hypocotyl?
4 Is there a distinct, peculiar shape to the cotyledons?
>  Pointed
>  Long and narrow
>  Oval.
5 Have the cotyledons peculiarities?
>  e.g. Backwardly directed lobes
>  Notched tips.

The order of the illustrations in the book follows this order of awareness.


**Notice too**

1 Is there an outstanding colour?
>  e.g. Bright green
>  Purplish hues.
2 Are the cotyledons or true leaves (or both) hairy?


The time of year can also help in identification; many weed seeds germinate during a limited season. Diagrams accompanying the sets of photographs show times of year at which germination is most likely to occur. These, and the extent of outbreaks, will be affected by weather and cultivations.

# Cotyledons: quick guide

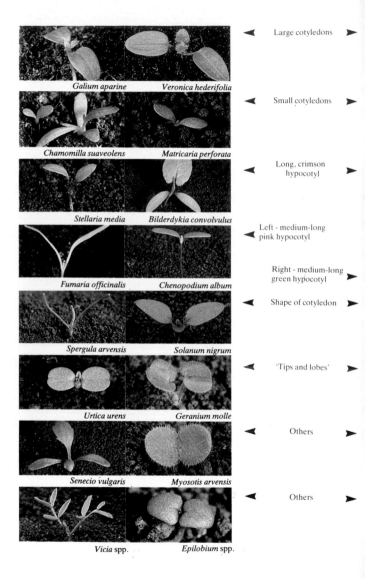

*Galium aparine* — *Veronica hederifolia*
◄ Large cotyledons ►

*Chamomilla suaveolens* — *Matricaria perforata*
◄ Small cotyledons ►

*Stellaria media* — *Bilderdykia convolvulus*
◄ Long, crimson hypocotyl ►

*Fumaria officinalis* — *Chenopodium album*
◄ Left - medium-long pink hypocotyl
Right - medium-long green hypocotyl ►

*Spergula arvensis* — *Solanum nigrum*
◄ Shape of cotyledon ►

*Urtica urens* — *Geranium molle*
◄ 'Tips and lobes' ►

*Senecio vulgaris* — *Myosotis arvensis*
◄ Others ►

*Vicia* spp. — *Epilobium* spp.
◄ Others ►

*Sinapis arvensis*      *Raphanus raphanistrum*     *Lapsana communis*

*Aphanes arvensis*     *Capsella bursa-pastoris*     *Papaver rhoeas*

*Polygonum aviculare*     *Polygonum persicaria*     *Polygonum lapathifolium*

*Atriplex patula*     *Chrysanthemum segetum*     *Thlaspi arvense*

*Veronica persica*     *Viola arvensis*     *Anagallis arvensis*

*Lamium amplexicaule*     *Lamium purpureum*     *Galeopsis tetrahit*

*Euphorbia peplus*     *Aethusa cynapium*     *Sonchus* spp.

*Cardamine hirsuta*     *Rumex* spp.     *Taraxacum officinale*

11

# Young plants: quick guide

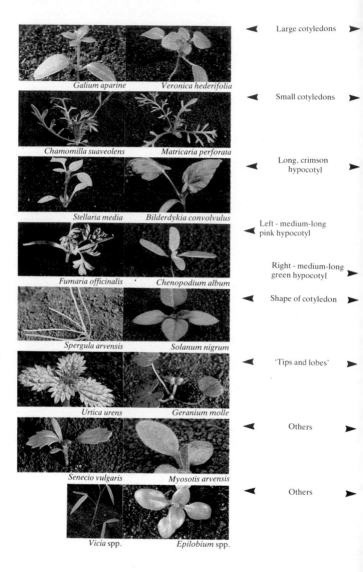

| | |
|---|---|
| *Galium aparine* | *Veronica hederifolia* |
| *Chamomilla suaveolens* | *Matricaria perforata* |
| *Stellaria media* | *Bilderdykia convolvulus* |
| *Fumaria officinalis* | *Chenopodium album* |
| *Spergula arvensis* | *Solanum nigrum* |
| *Urtica urens* | *Geranium molle* |
| *Senecio vulgaris* | *Myosotis arvensis* |
| *Vicia* spp. | *Epilobium* spp. |

◄ Large cotyledons ►

◄ Small cotyledons ►

◄ Long, crimson hypocotyl ►

◄ Left - medium-long pink hypocotyl

Right - medium-long green hypocotyl ►

◄ Shape of cotyledon ►

◄ 'Tips and lobes' ►

◄ Others ►

◄ Others ►

*Sinapis arvensis*     *Raphanus raphanistrum*     *Lapsana communis*

*Aphanes arvensis*     *Capsella bursa-pastoris*     *Papaver rhoeas*

*Polygonum aviculare*     *Polygonum persicaria*     *Polygonum lapathifolium*

*Atriplex patula*     *Chrysanthemum segetum*     *Thlaspi arvense*

*Veronica persica*     *Viola arvensis*     *Anagallis arvensis*

*Lamium amplexicaule*     *Lamium purpureum*     *Galeopsis tetrahit*

*Euphorbia peplus*     *Aethusa cynapium*     *Sonchus* spp.

*Cardamine hirsuta*     *Rumex* spp.     *Taraxacum officinale*

13

## *Galium aparine*
# Cleavers
Goose Grass
Herrif
Sticky Willie

1. **Cotyledons**: large, oblong, notched at end (**1a**); dark green, sometimes purplish
2. **Stem**: square and with hooked spines
3. **First true leaves**: in whorls of four
4. **Young buds** develop early in the axils of the cotyledons

A common annual weed of arable land and waste places

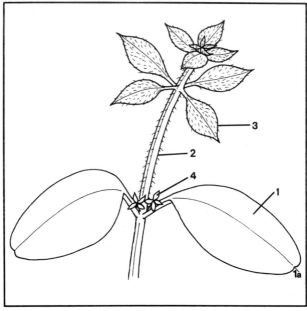

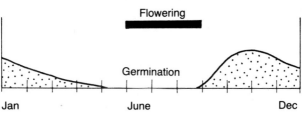

# *Veronica hederifolia*
# Ivy-leaved Speedwell

1. **Cotyledons**: <u>very large</u>, often purple below,
   <u>on long curved stalks</u>
   **Hypocotyl**: medium-long
2. **First true leaves**: <u>hairy</u>, toothed near the base
   **Similar weed**: cleavers: distinguished from cleavers by
   the absence of a notch in the end of the cotyledon

A common annual weed of arable land and gardens, particularly where fertility is good. The seedlings are obvious
during the winter

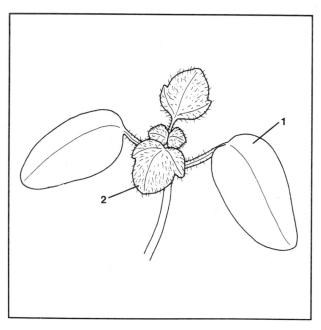

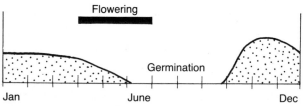

## *Sinapis arvensis*
## Charlock
### Wild Mustard

1. **Cotyledons**: <u>large, kidney shaped</u>. Medium length, nearly vertical cotyledon stalks carry the blades of the cotyledon well away from the ground
   **Hypocotyl**: short
2. **First true leaves**: regular indentations of first true leaf; broad, rounded tip
   **Similar weeds**: charlock is difficult to distinguish from white mustard, black mustard and runch (but compare germination period with that of runch)

A very common weed of arable and waste land

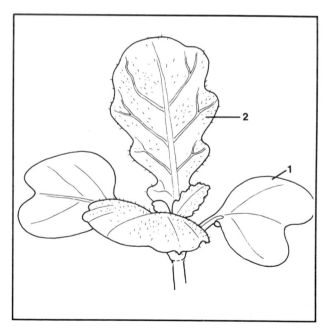

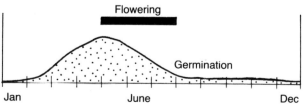

## *Raphanus raphanistrum*
## Runch
Wild Radish
White Charlock

1. **Cotyledons**: <u>large, kidney shaped</u>, medium length,
   cotyledon stalks carry the cotyledon blades away
   from the ground, purplish hue
   **Hypocotyl**: <u>short, purple</u>
2. **First true leaves**: indented, rough
   **Later leaves**: have several independent lobes at the base
   **Similar weed**: charlock–the purplish hue and roughness
   of runch distinguishes it

A common annual weed of arable and waste land

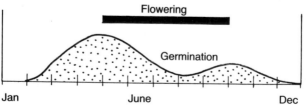

21

## *Lapsana communis*
## Nipplewort

1. **Cotyledons**: <u>oval</u>, carried off the ground
2. **First true leaves**: <u>hairy</u> with wavy, irregular outline,
   yellowish-green
   **Later leaves**: pronounced, irregular outline: early
   leaves in rosette, near the ground. True leaves
   appear singly – not in pairs

An occasional annual weed of arable land, but common in
gardens and waste land

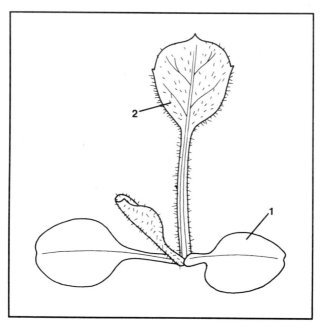

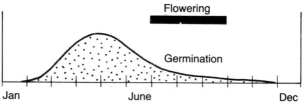

23

## *Chamomilla suaveolens*
## *(Matricaria matricarioides)*
# Pineappleweed
### Rayless Mayweed

1. **Cotyledons**: <u>very small</u>, appear close to the ground
2. **First true leaves**: <u>narrow with few lobes</u> (often two)
   **Later leaves**: narrow, with numerous lobes, shiny
   **Similar weeds**: other mayweeds. First true leaves of pineappleweed have fewer lobes, and the stalks of later leaves are broader than those of other mayweeds

A common annual weed of arable land, particularly in gateways and on headlands

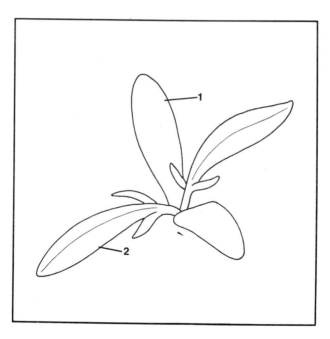

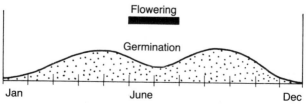

Flowering

Germination

Jan        June        Dec

24

## *Matricaria perforata*
## Scentless Mayweed

1. **Cotyledons**: <u>very small</u>; they appear close to the ground
2. **First true leaves**: <u>narrow</u>; the terminal lobes tend (in comparison with those of other mayweeds) to be shorter and broader
3. **Later leaves**: narrow, with numerous lobes, shiny
   **Similar weeds**: other mayweeds

A common and important annual weed of arable land

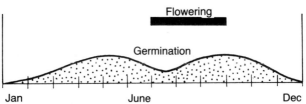

## *Aphanes arvensis*
## Parsley-piert

1. **Cotyledons**: <u>small</u>, hairless
   **Hypocotyl**: short/medium-short
2. **First true leaves**: bright bluish-green, three-lobed,
   roughly hairy on the margins
3. **Later leaves**: spreading; each of the three lobes is
   further divided into 5–7 parts
4. **Leaf stipules**: fused into leaf-like cup

A common annual weed of arable land

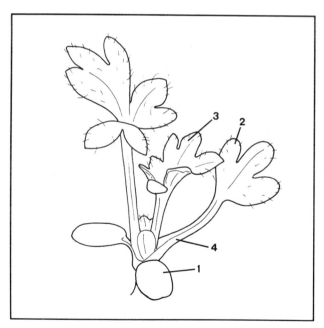

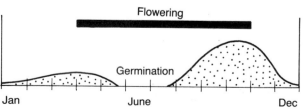

Flowering

Germination

Jan          June          Dec

# *Capsella bursa-pastoris*
# Shepherd's-purse

1. **Cotyledons**: <u>small, narrow</u>
2. **First true leaves**: grey-green with <u>star-like hairs</u>;
   elliptical with distinctive petioles (**2a**)
   **Hypocotyl**: short
   **Later leaves**: very variable, usually much divided,
   forming a rosette
   **Similar weeds**: shepherd's purse, because it is so
   variable, may resemble several other species

A common weed of arable land, waste land and gardens

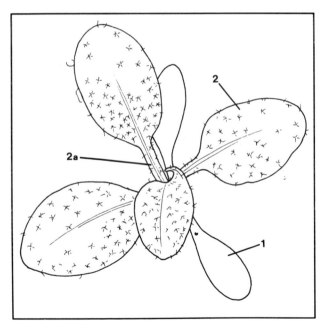

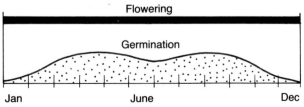

## *Papaver rhoeas*
## Common Poppy
### Field Poppy

1. **Cotyledons**: <u>small, narrow, pointed</u>; without petioles; bluish-green
2. **First true leaves**: <u>entire, bluish-green</u> with small simple hairs
   **Hypocotyl**: short
   **Later leaves**: indentations are apparent; older leaves are deeply divided
   **Similar seedlings**: shepherd's purse p 30–cotyledons distinguish. Other species of *Papaver*

A common weed of arable land and waste places

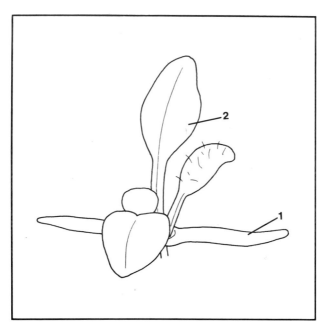

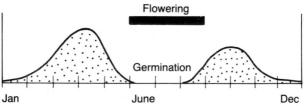

Flowering

Germination

Jan                    June                    Dec

## *Stellaria media*
## Common Chickweed

1. **Cotyledons**: oval, pointed, light-coloured tip, bright green
2. **First true leaves**: entire, hairy
3. **Hypocotyl**: long, red-purple

A very common annual weed of arable land and gardens. The weed chokes other plants as it spreads

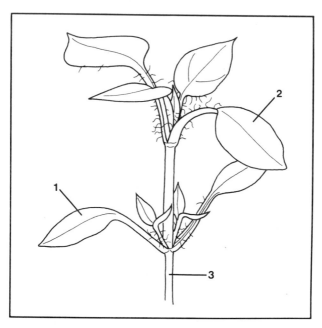

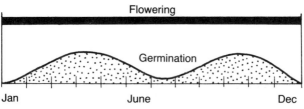

## *Bilderdykia (Fallopia) convolvulus*
## Black-bindweed

1. **Hypocotyl**: <u>long, crimson</u>
2. **Cotyledons**: <u>long</u>, reddish-green
3. **First true leaves**: heart-shaped <u>with rounded lobes</u> at the base, reddish-green and shiny
4. **Later leaves**: tightly rolled
   **Similar weeds**: fat-hen and orache have similar (but smaller) cotyledons. Their mealy first leaves distinguish them from black-bindweed

A common annual weed of arable land and gardens

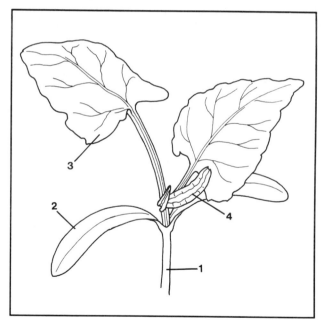

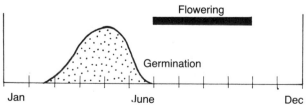

Flowering

Germination

Jan          June          Dec

## *Polygonum aviculare*
## Knotgrass

1. **Hypocotyl**: <u>long, crimson</u>
2. **Cotyledons**: <u>long and narrow, thick-looking</u>; they are set at 40° angles (in elevation) and also are at an angle in plan (they are not directly opposite, **2a**)
3. **First true leaves**: <u>broad</u>
   **Similar weeds**: fat-hen. Knotgrass lacks the mealy surface of cotyledons and leaves. Its cotyledons are set at angles, where those of fat-hen are opposite and parallel to the ground.
   Pale persicaria and redshank. The narrow cotyledons of knotgrass distinguish it

A very important weed of arable land. It occurs in waste places, on paths and in gardens

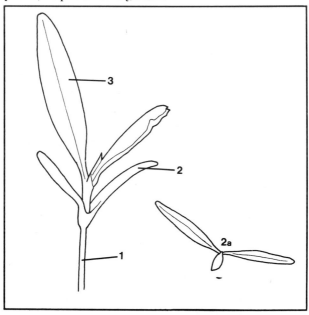

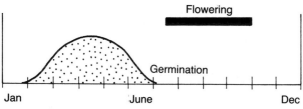

## *Polygonum persicaria*
## Redshank
Persicaria

1. **Hypocotyl**: <u>long, bright scarlet</u>
2. **Cotyledons**: <u>oval</u>, dark green tinged with red
3. **First true leaves**: <u>broad</u> with few hairs
4. **Later leaves**: red blotch
   **Similar weed**: pale persicaria – has narrow first true leaves with hairs

A common annual weed of arable and waste lands, particularly where they lie damp

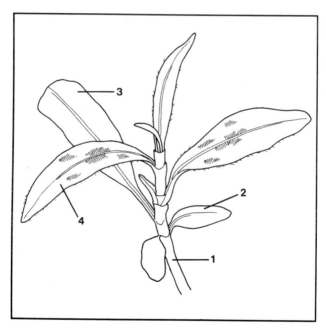

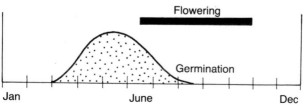

Flowering

Germination

Jan          June          Dec

## *Polygonum lapathifolium*
## Pale Persicaria
Willow Weed

1. **Hypocotyl**: <u>long, bright scarlet</u>
2. **Cotyledons**; oval – with rounded tips
3. **First true leaves**: noticeably <u>silver with hairs</u>
   **Similar weed**: redshank–lacks the silvery hairs on the first true leaves

A common annual weed of cultivated land especially on moist soils

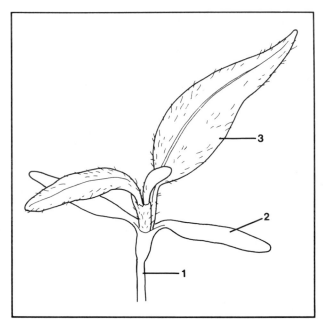

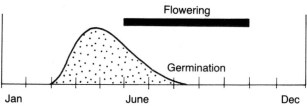

# *Fumaria officinalis*
# Common Fumitory

1. **Cotyledons**: very long and narrow, bluish-green
2. **Hypocotyl**: long, pinkish
3. **First true leaves**: much divided, bluish-green

A common annual weed of arable land and gardens, part-icularly on light soils and in drier, eastern regions

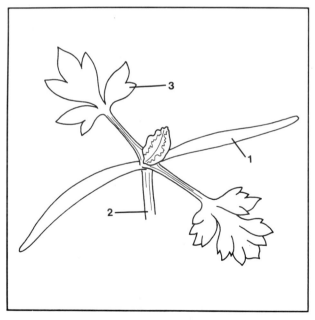

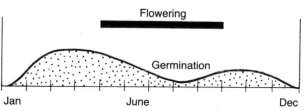

Flowering

Germination

Jan          June          Dec

45

## *Chenopodium album*
## Fat-hen
Muck-weed

1. **Hypocotyl**: slender, medium-long (1 cm), purplish
2. **Cotyledons**: much longer than wide, purplish on the underside
3. **First true leaves**: mealy
   **Later leaves**: shape is variable
   **Similar weed**: orache. Orache seedlings are bigger (fatter looking) and lack the purplish colour on both hypocotyl and the undersides of cotyledons and leaves

A very common annual weed of arable land and gardens

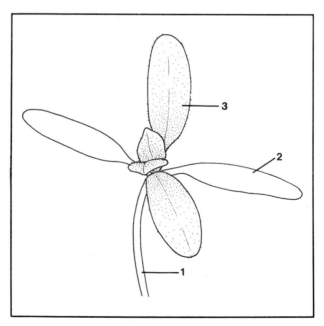

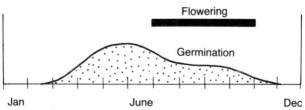

Flowering

Germination

Jan          June          Dec

## *Atriplex patula*
## Common Orache
Creeping Fat-hen

1. **Hypocotyl**: <u>long</u> – pinkish green
2. **Cotyledons**: <u>long, narrow, mealy</u>, pale green
3. **First true leaves**: mealy, green with little purple colouring
   **Leaf bud**: <u>mealy</u>
   **Later leaves**: variable in shape, toothed
   **Similar weed**: fat-hen. This is more slender; the cotyledons and first leaves appear thinner than those of orache. Orache lacks the purple colours of fat-hen seedlings

A common annual weed of arable and waste land

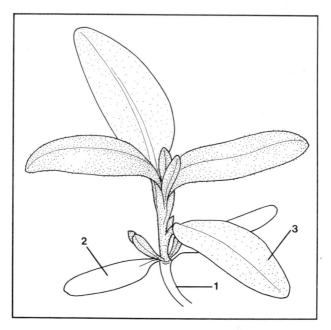

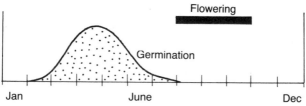

Flowering

Germination

Jan | June | Dec

48

# *Chrysanthemum segetum*
# Corn Marigold

1. **Hypocotyl**: <u>long</u>, light bluish-green
2. **Cotyledons**: medium size, oval, carried well from the ground
3. **First true leaves**: narrow with teeth (like mayweeds) but variable. Waxy and bluish-green

An annual weed growing almost exclusively in arable fields on light, acid soils.

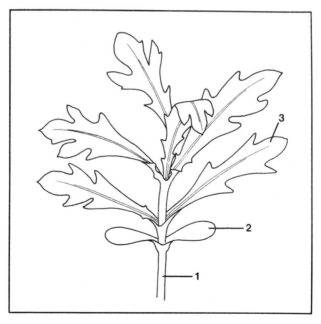

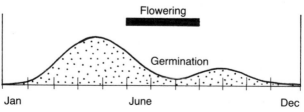

Flowering

Germination

Jan       June       Dec

50

## *Thlaspi arvense*
## Field Penny-cress

1. **Cotyledons**: medium size, <u>light green</u>, broad-oval, with long stalks. The tip often curves downwards and quickly becomes discoloured
   **Hypocotyl**: short
2. **First true leaves**: hairless, <u>light green</u>, with slightly wavy margins
3. **Later leaves**: slightly wavy margins; when crushed give an unpleasant smell

A common annual weed of arable land and waste places

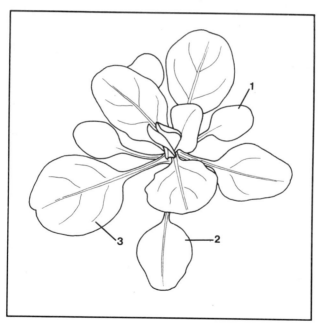

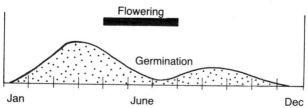

## *Spergula arvensis*
## Corn Spurrey

1. **Cotyledons**: very long and very narrow
2. **First true leaves**: very long and very narrow yet fleshy;
   grass green
   **Similar weeds**: may superficially be mistaken for
   annual meadow-grass

A weed of arable land, particularly on light soils which are
acidic. A local rather than widespread annual weed – most
common in north and west England

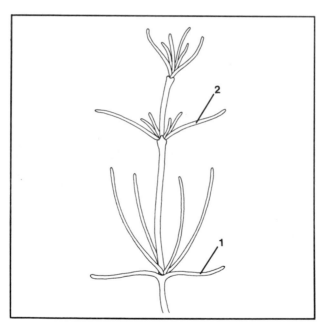

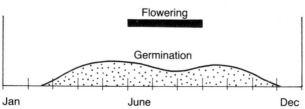

Flowering

Germination

Jan            June            Dec

# *Solanum nigrum*
# Black Nightshade

1. **Cotyledons**: <u>sharply pointed</u>, hairy, often purplish
   **Hypocotyl**: short, but cotyledons are carried above the ground; often purplish
2. **First true leaves**: entire, hairy, <u>dull dark green, tinged with blue or purple</u>

A common annual weed of arable land and gardens

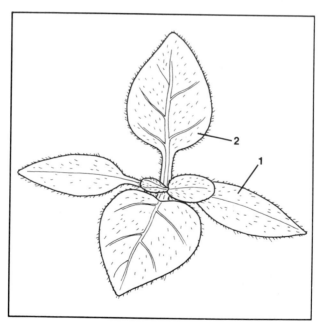

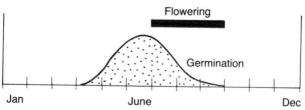

Flowering

Germination

Jan          June          Dec

57

# *Veronica persica*
# Common Field-speedwell

1. **Cotyledons**: spade-shaped, medium–large size
2. **First true leaves**: margins shallowly and regularly notched. <u>Hairy</u>, often greyish-green
   **Similar weeds**: other *Veronica* species (except *V. hederifolia*). Seedlings of common field-speedwell are larger

A common annual weed of arable land and gardens

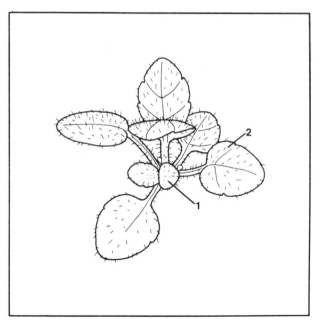

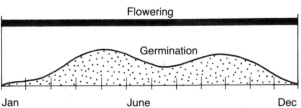

Flowering

Germination

Jan                    June                    Dec

# *Viola arvensis*
# Field Pansy

1. **Cotyledons**: oblong, dark green; indentation at the tip
2. **First true leaves**: broadly-rounded apex
3. **Later leaves**: broad, with lobes, slightly hairy; the leaf stalks are noticeable and are hairy. Leaves are initially rolled and open one by one

A common annual weed of arable land

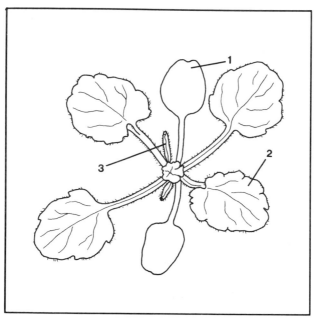

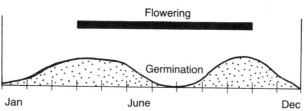

# *Anagallis arvensis*
# Scarlet Pimpernel

1. **Cotyledons**: <u>small</u>, appear close to the ground, shiny <u>dark green, triangular</u> and pointed
2. **First true leaves**: shiny, triangular; <u>dark spots on the underside</u>
3. **Later leaves**: triangular, shiny, hairless
   **Similar weed**: the cotyledons have a similar shape to those of chickweed. Chickweed seedlings, however, have a long hypocotyl and are much lighter green

An annual weed of cultivated land and waste places

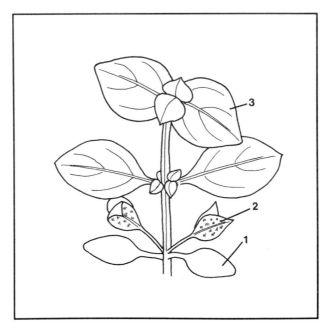

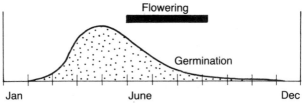

Flowering

Germination

Jan                June                Dec

## *Urtica urens*
## Small Nettle

1. **Cotyledons**: medium size, appear close to the
   ground, hairy, <u>notched at the apex</u>
2. **First true leaves**: pointed teeth, with stinging hairs on the
   upper surface
   **Similar weed**: perennial stinging nettle has similar
   seedlings but these have shorter cotyledons than
   the small nettle, and less pointed teeth on the first
   leaves

A very common annual weed of arable land and gardens

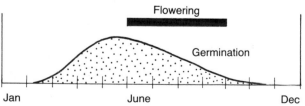

## *Geranium molle*
## Dove's-foot Crane's-bill

1. **Cotyledons**: breadth exceeds width; asymmetrical; borne on <u>long</u>, hairy stalks
2. **First true leaves**: much divided. Each true leaf develops singly
   **Similar weeds**: meadow crane's-bill and cut-leaved crane's-bill

A common annual weed of arable land and gardens

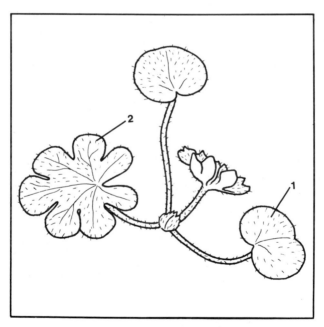

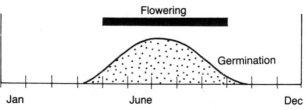

67

## *Lamium amplexicaule*
## Henbit Dead-nettle

1. **Cotyledons**: the nearly vertical stalks of the cotyledons make this seedling stand away from the ground. The cotyledons are almost horizontal but the margins tend to be directed downward
2. **Pronounced notch at the stalk**
3. **First true leaves**: evenly notched, with prominent, branched lobes
   **Similar weed**: red dead-nettle. Henbit is paler

A common annual weed of arable land

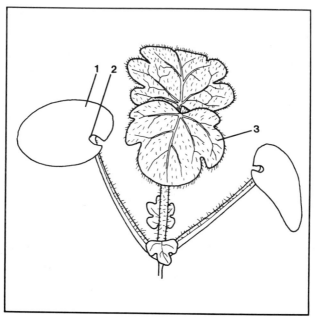

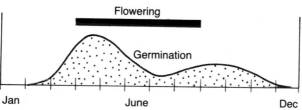

69

## *Lamium purpureum*
## Red Dead-nettle

1. **Cotyledons**: nearly vertical cotyledon stalks and horizontal blades make this seedling stand away from the ground
2. **Pronounced notch at the stalk**
3. **First true leaves**: evenly notched, with prominent, branched veins, hairy
   **Similar weed**: it is difficult to distinguish from henbit but is of a darker green. The cotyledons of red dead-nettle appear more rigid and horizontal

A widespread annual weed of arable land and waste grass-land; especially common in gardens

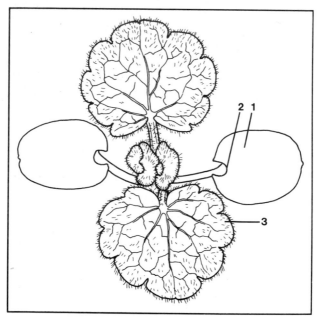

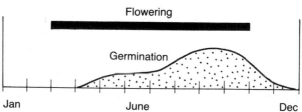

Flowering

Germination

Jan        June        Dec

## *Galeopsis tetrahit*
## Common Hemp-nettle
Day-nettle

1. **Cotyledons**: fairly large, oval with backwardly directed lobes at the base of the leaf blades
2. **First true leaves**: have regular marginal teeth and veining

A common annual weed of arable land

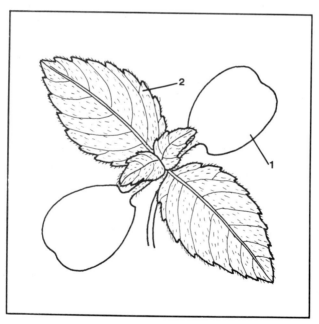

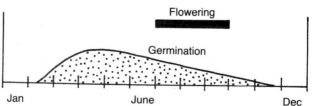

*Senecio vulgaris*
Groundsel

**Hypocotyl**: purplish, medium – carries the cotyledons just above the ground
1. **Cotyledons**: <u>narrow</u>; often purplish underneath
2. **First true leaves**: <u>step-like teeth</u>
   **Later leaves**: variable, but always indented or with teeth. May be with or without hairs

A common annual weed in most places

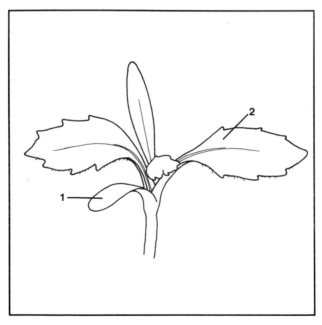

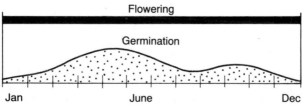

Flowering

Germination

Jan                    June                    Dec

75

## *Myosotis arvensis*
## Field Forget-me-not

1. **Cotyledons**: small, hairy, dark green
   **Hypocotyl**: short; the seedling appears close to the ground
2. **First true leaves**: entire, hairy; open one by one, not in pairs

An annual weed occuring on arable and waste land

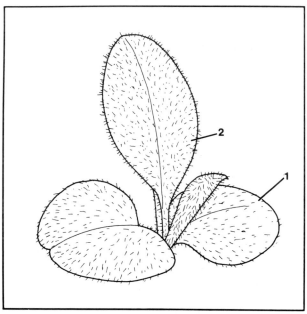

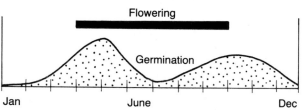

Flowering

Germination

Jan          June          Dec

## *Euphorbia peplus*
## Petty Spurge

1. **Cotyledons**: oval to long carried on medium length hypocotyl
2. **First true leaves**: entire and carried on a short stem above the cotyledons

Spurges contain a milky juice which quickly exudes when the stem is broken. Petty spurge is a common annual weed of gardens: sun spurge, with similar seedling characteristics is common on arable land.

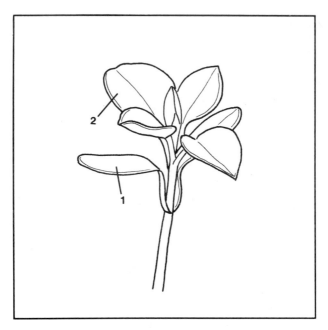

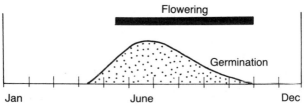

Flowering

Germination

Jan    June    Dec

# *Aethusa cynapium*
## Fool's Parsley

1. **Cotyledons**: length exceeds width; gradually narrowing into the stalks; shiny
   **Hypocotyl**: medium – carrying the cotyledon above the ground
2. **First true leaves**: deeply divided into three lobes, dark green, hairless

A common weed of arable land. It is annual or biennial

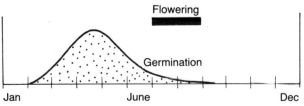

# *Sonchus* spp.
## Sow-thistles

1. **Cotyledons**: oval, medium-sized, close to the ground
2. **First true leaves**: dull bluish-green above, purple hues beneath
3. downwardly directed teeth on the margin
4. few, thick white hairs
   **Similar weeds**: Seedlings of the various sow-thistle species are difficult to separate. *S. arvensis* (perennial sow-thistle), *S. asper* and *S. oleraceus* are similar

Common weeds of arable land and gardens

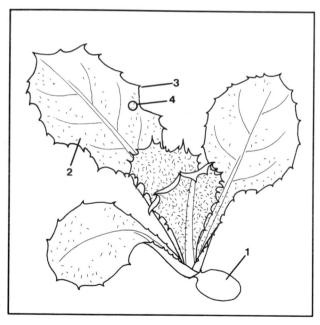

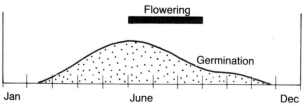

<table>
<tr><td>Cotyledons remain below ground</td><td>

*Vicia sativa*
# Common Vetch
*Vicia hirsuta*
# Hairy Tare
</td></tr>
</table>

*Vicia sativa*
# Common Vetch
*Vicia hirsuta*
# Hairy Tare

The cotyledons of tares and vetches remain below ground
First true leaves have two leaflets *(V. sativa)*; or four leaflets in opposite pairs *(V. hirsuta)*
Both are annual weeds of arable land throughout the British Isles and can be common in some areas

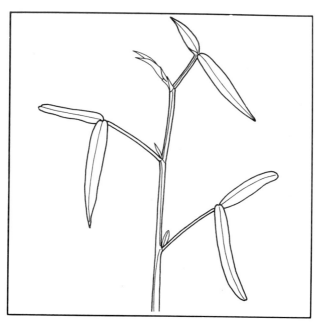

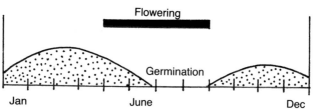

Flowering

Germination

Jan        June        Dec

*Epilobium* spp.
Willowherbs

1. **Cotyledons**: <u>small</u>, entire
   **Hypocotyl**: very short; the young seedling lies close to the ground
2. **First and later true leaves:** very shiny, yellowish-green, lying close to the ground

Seeds of willowherbs are dispersed over long distances by wind and seedlings occur in waste places, nurseries and gardens

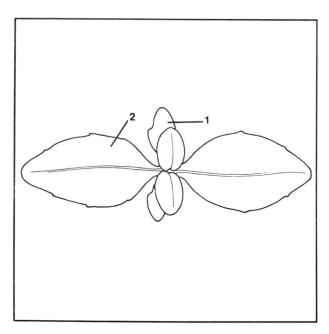

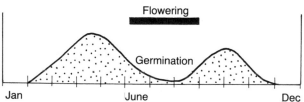

Flowering

Germination

Jan    June    Dec

## *Cardamine hirsuta*
## Hairy Bitter-cress

1. **Cotyledons**: oval
2. **First true leaves**: kidney shaped, hairy
3. **Later leaves**: long stalks with pairs of leaflets

A common annual weed of gardens and nurseries, especially in container-grown nursery stock

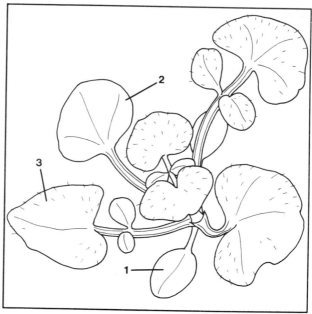

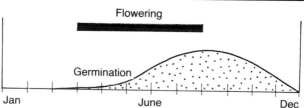

Flowering

Germination

Jan    June    Dec

## *Rumex* spp.
## Docks

1. **Cotyledons**: narrow, medium size, diamond-shaped
   **Hypocotyl**: short, but cotyledons carried above the ground
2. **First true leaves**: tinged with purple; initially are rolled, with a frill

Species of docks are difficult to distinguish from each other in the seedling stage. Many are common perennial weeds of grassland and waste places

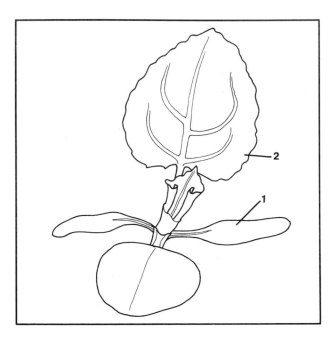

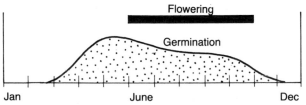

Flowering

Germination

Jan                    June                    Dec

# *Taraxacum officinale*
# Common Dandelion

1. **Cotyledons**: oval
2. **First true leaves**: dark, shiny green and hairless; margins have backwardly (downwardly) directed teeth
   **Similar weeds**: perennial sow-thistles also have backwardly directed teeth but are bluish-green

A common perennial weed of grassland and gardens

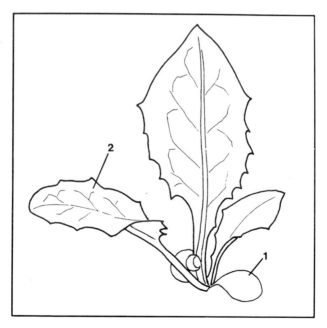

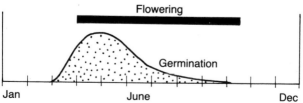

Flowering

Germination

Jan    June    Dec

# Glossary

**Annual**
A plant that completes its life cycle within 12 months of seed germination.

**Apex**
(of a leaf)
The top of the leaf — the point furthest away from the stem.

**Axil**
The upper angle between a leaf (or its petiole) and the stem from which it arises.

**Biennial**
A plant that completes its life cycle within 2 years of seed germination (but not within one year).

**Cotyledons**
The first leaves of a seedling, present within the seed. These often rise above the ground and become green; they are usually different from the first true leaves. Occasionally the cotyledons remain beneath the ground.

**Entire**
Without cuts or teeth (of a leaf).

**Hypocotyl**
The stem supporting the cotyledons.

**Lobe**
A roundish part of a leaf — divided from the main part but not separate as is a leaflet.

**Margin**
The edge or border of a leaf.

**Mealy**
Description of a leaf which appears spotted with or covered by a white powder.

**Node**
The place on the stem from which a leaf (or its petiole) arises.

**Petiole**
The leaf stalk.

**Rosette**
A cluster of leaves which resembles a rose, usually near the ground.

**Stipule**
A scale or leaf-like appendage at the base of a petiole.

**True leaves**
The later leaves of a seedling, developing after the cotyledons.

**Whorl**
Three or more leaves arising at the node.